Words to Know

arms

clams

color

eggs

starfish

This is a **starfish**.

UNDER THE SEA ANIMALS

STARFISH

Amy Culliford

TABLE OF CONTENTS

A Pelican Book

SEAHORSE PUBLISHING

Teaching Tips for Caregivers and Teachers:

Research shows that one of the best ways for students to learn a new topic is to read about it.

Before Reading

- Read the title and predict what the book will be about.
- Read the "Words to Know" and discuss the meaning of each word.
- Read the back cover to see what the book is about.

During Reading

- When a student gets to a word that is unknown, ask them to look at the rest of the sentence to find clues to help with the meaning of the unknown word.
- Motivate students with praise and encouragement.

After Reading

- Discuss the main idea of the book.
- Ask students to give one detail that they learned in the book.

Sight Words

a
all
and
any
are
be
big
can
eat
have
is
lay
little
long
many
some
this

starfish

A starfish can be any **color**.

color

Some starfish are big, and some are little.

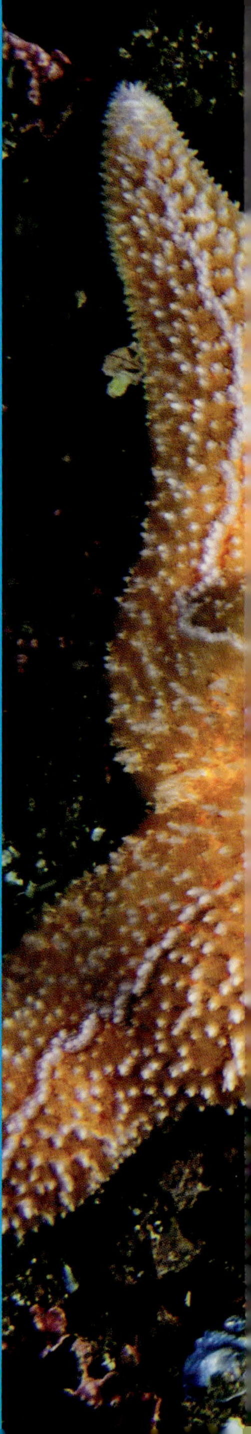

Many starfish have long **arms**.

arms

Some starfish eat **clams**.

clams

All starfish lay **eggs**.

eggs

Index

Written by: Amy Culliford
Design by: Under the Oaks Media
Series Development: James Earley
Editor: Kim Thompson

Photos: Ethan Daniels: cover; Efired: p. 4-5; Dallas Reeves: p. 7; RLS Photo: p. 9; Adamke: p. 11; Pooh & Ball: p. 13; Pablo Henrique Poblete: p. 15

Library of Congress PCN Data
Starfish / Amy Culliford
Under the Sea Animals
ISBN 978-1-63897-067-5 (hard cover)
ISBN 978-1-63897-153-5 (paperback)
ISBN 978-1-63897-239-6 (EPUB)
ISBN 978-1-63897-325-6 (eBook)
Library of Congress Control Number: 2021945248

Printed in the United States of America.

Seahorse Publishing Company
www.seahorsepub.com

Published in the United States
Seahorse Publishing
PO Box 771325
Coral Springs, FL 33077